Prepper's Water Survival Bible

How to Find, Purify, and Store Water to Keep You and Your Family Safe and Healthy During an Emergency.

NEVADA BECKER

1

Table of content

INTRODUCTION

Welcome and Purpose

In the realm of emergency preparedness, the significance of water cannot be overstated. As we embark on this journey through the Prepper's Water Survival Bible, our foremost objective is to guide individuals and families towards a heightened awareness of the pivotal role water plays in times of crisis. This compendium is not just a manual but a lifeline, providing indispensable insights into securing a clean and sustainable water supply during emergencies. So, welcome to a comprehensive exploration of water survival strategies that will empower you to safeguard the well-being of yourself and your loved ones.

Importance of Water in Emergency Preparedness

The core of emergency preparedness rests on the foundational understanding of the vital role water plays in sustaining life. Often underestimated, water is the linchpin that can determine survival or peril during unforeseen circumstances. Its significance extends beyond mere hydration; water is the linchpin of health, sanitation, and sustenance. In this chapter, we delve into the intricate web of water's importance, uncovering its multifaceted contributions to human well-being and survival. From the physiological requirements of the body to the broader implications for community resilience, the exploration aims to instill a profound appreciation for the life-sustaining essence of water.

Understanding Water Sources and Risks

Before navigating the labyrinth of water survival, it is imperative to comprehend the diverse sources from which this elixir can be drawn and the lurking risks that may compromise its purity. Natural water bodies, wells, and municipal supplies present distinct challenges and vulnerabilities. This chapter serves as a compass, guiding readers through the intricacies of identifying, evaluating, and mitigating risks associated with different water sources. By unraveling the complexities of contamination, scarcity, and the dynamic nature of water availability, we equip individuals with the knowledge needed to make informed decisions when sourcing and managing water in emergency scenarios.

As we journey through the following chapters, each segment unfolds like a chapter in a suspenseful novel, revealing crucial elements that collectively contribute to the overarching narrative of water survival. The book, in its entirety, is a roadmap towards self-reliance and community resilience, painting a vivid picture of a world where water, once seen as commonplace, becomes the cornerstone of preparedness and survival.

Chapter 1

Assessing Your Water Needs

In the intricate tapestry of emergency preparedness, understanding and assessing one's water needs is the foundational stitch. As we embark on this chapter, the first brushstroke on the canvas of water survival involves a meticulous examination of the factors influencing our daily water requirements. Calculating the volume of water needed per person per day is not merely a task but a crucial skill that can tip the scales between sufficiency and scarcity in times of crisis.

Calculating Daily Water Requirements

At the heart of water preparedness lies the fundamental need to calculate daily water requirements. This calculation is not a one-size-fits-all endeavor; it demands an intimate understanding of individual habits, local climate, and health conditions. It extends beyond mere hydration needs, considering factors such as cooking, hygiene, and potential medical requirements. Delving into this intricacy ensures a personalized and accurate approach, offering a bespoke roadmap to building a resilient water supply tailored to the unique needs of each household.

Considering Specific Needs for Children, Elderly, and Pets

As we navigate the landscape of water needs, it is imperative to recognize the nuances associated with specific demographics within a household. Children, being more susceptible to dehydration, necessitate additional considerations in water planning. Similarly, the elderly may have unique health requirements, elevating their need for clean water. Factor in the four-legged members of the family, and the complexity deepens. This section is a compass for tailoring water strategies to encompass the diverse needs of every family member, ensuring a holistic and inclusive approach to water preparedness.

Estimating Water Storage Duration

Once the daily water needs are meticulously calculated, the next dimension of preparedness unfolds in estimating the duration for which a water supply must be sustained. This facet is a strategic layer, considering potential disruptions and the time it might take for normalcy to be restored. By integrating knowledge of local emergency response capabilities and historical data, individuals can gauge the necessary duration for which their water storage should be fortified. This forward-looking perspective transforms water preparedness from a static reservoir to a dynamic strategy, capable of adapting to the ebb and flow of crisis scenarios.

Chapter 2

Water Sources in Your Area

The journey into water survival extends beyond personal needs to a broader exploration of the local landscape. Chapter 2 unfurls like a map, guiding individuals in identifying and harnessing the diverse water sources in their vicinity. This chapter is not just about recognizing the obvious streams or lakes but delves into the intricacies of the local water tapestry, unraveling the potential lifelines hidden in plain sight.

Identifying Local Water Sources

A critical step in water preparedness is acquainting oneself with the local water sources. Beyond the tap or the bottled water aisle, a community often has access to a myriad of water bodies. From rivers and ponds to underground aquifers, the possibilities are diverse. This section is a compass, leading readers through a process of discovery, helping them recognize and catalog the various water sources within their reach. In doing so, it transforms a seemingly mundane landscape into a potential reservoir of life-sustaining liquid.

Evaluating the Safety of Natural Water Bodies

While identifying local water sources is the first step, ensuring their safety is the linchpin of this chapter. Natural water bodies can be both a blessing and a potential hazard. Understanding the factors that influence water quality – from industrial runoffs to agricultural pollutants – becomes paramount. This segment equips individuals with the knowledge to evaluate the safety of these sources, providing insights into testing methods and red flags that might compromise the purity of the water. In essence, it transforms the act of drawing water from a nearby stream into a calculated decision based on a nuanced understanding of environmental factors.

Mapping Nearby Water Access Points

Navigating the local water landscape requires more than just mental recognition; it demands a tangible map. This section unfolds like a cartographer's guide, encouraging individuals to map out the nearby water access points systematically. This mapping is not just a logistical exercise but a strategic one, considering factors such as accessibility, safety, and potential contaminants. It transforms the act of sourcing water into a deliberate journey, where individuals are not merely reacting to a crisis but proactively charting a course through their local water topography.

Chapter 3

Water Purification Methods

Water, the elixir of life, holds the promise of sustenance but can also harbor hidden threats. Chapter 3 unfurls like a handbook of alchemy, delving into the myriad methods of transforming potentially contaminated water into a safe and life-sustaining resource. This chapter is not just a compendium of techniques; it is a guide through the labyrinth of purification methods, equipping individuals with the knowledge to turn crisis into opportunity.

Boiling Techniques and Guidelines

The most ancient and elemental of water purification methods, boiling stands as a sentinel against waterborne pathogens. This section delves into the art and science of boiling, going beyond the mere bubbling of water. It unfolds like a culinary masterclass, elucidating the optimal temperatures, durations, and potential pitfalls of boiling as a purification method. From the subtle dance of water particles to the microbial battlefield within, readers are immersed in the transformative power of this time-tested technique.

Filtration Systems and Options

In the modern tapestry of water purification, filtration systems emerge as sophisticated guardians, sifting out impurities with a precision that rivals nature itself. This segment navigates through the intricate world of filtration, exploring options ranging from basic ceramic filters to advanced micron-level purifiers. It's not just about selecting a filter; it's about understanding the unique challenges posed by different contaminants and tailoring filtration solutions to the specific needs of the water source. Like a connoisseur of fine wines, individuals are guided through the nuances of selecting the perfect filtration system that complements the character of their water.

Chemical Treatment and Disinfection

Chemical treatment stands as a sentinel in the arsenal of water purification, introducing an element of controlled alchemy to render water safe. This section unfolds like a laboratory manual, demystifying the chemistry behind disinfectants such as chlorine and iodine. It explores not only the how but also the why, delving into the mechanisms by which these chemicals neutralize pathogens. From tablets to liquid solutions, individuals are equipped with the knowledge to wield chemical disinfection as a precise and effective tool in their water survival kit.

Solar Water Disinfection (SODIS) Method

As the sun ascends in the sky, a natural purifier takes center stage – solar radiation. The Solar Water Disinfection (SODIS) method harnesses the sun's energy to cleanse water from pathogens. This segment is a journey into the realm of harnessing sunlight as a purification powerhouse. From the science behind UV rays to the practicalities of using transparent containers, individuals are guided through the intricacies of leveraging this free and abundant resource to transform questionable water into a pristine elixir.

Chapter 4

DIY Water Collection and Storage

In the symphony of water preparedness, this chapter unfolds like a craftsman's manual, guiding individuals in the art of DIY water collection and storage. This chapter transcends the conventional approach of relying solely on pre-packaged solutions, empowering readers to become architects of their water resilience. It's not just about storing water; it's about crafting a bespoke system that harmonizes with the unique contours of their surroundings.

Rainwater Harvesting Systems

Nature provides a bountiful source of pure water straight from the heavens – rain. Rainwater harvesting systems stand as a testament to the synergy between human ingenuity and natural abundance. This section unfurls like an engineer's blueprint, guiding individuals in capturing and utilizing rainwater effectively. From the design of rooftop collection systems to the intricacies of storage tanks, readers are immersed in the craftsmanship of harnessing the sky's gift for their water needs. It transforms the act of waiting for rain into an active engagement with nature, where every drop becomes a precious resource.

Constructing Water Barrels and Containers

The vessels that cradle our water play a pivotal role in preserving its purity. This segment unfolds like a carpenter's guide, exploring the materials, designs, and construction techniques for water barrels and containers. It is not just about having a receptacle; it's about creating a sanctuary where water remains untainted and ready for use. From repurposed barrels to specially designed containers, readers are guided through the nuances of constructing their water reservoirs, turning storage into a craft where form meets function.

Tips for Storing Water Safely

Storage is not a passive act; it's a dynamic interplay between container and content. This section serves as a custodian's handbook, offering insights into the delicate dance between storage materials, location, and maintenance. From the choice of containers to the ideal storage conditions, readers are equipped with the knowledge to safeguard their water against the subtle threats that may compromise its purity. It's not just about amassing water; it's about curating a reservoir that stands resilient against the test of time and external influences.

Rotating Water Storage Supplies

The temporality of water storage introduces an additional layer of strategy – rotation. Like a choreographer orchestrating a dance, this section guides individuals in the rhythmic rotation of their water supplies. It's not just about amassing gallons; it's about ensuring a dynamic reservoir where the oldest water is always the first to be used. From the logistics of rotating containers to the discipline of maintaining a constant supply, readers are immersed in the art of water storage as a cyclical and renewable resource.

Chapter 5

Emergency Water Strategies

As the narrative of water survival unfolds, Chapter 5
emerges as a dynamic exploration of emergency water
strategies – a compendium of inventive solutions to quench
thirst when conventional sources falter. It is a chapter that
transcends the boundaries of the expected, guiding
individuals through the urban jungle, the botanical realm,
and the compact convenience of emergency water packs.
Like a survivalist's handbook, this chapter equips readers
with the skills to turn adversity into opportunity,
transforming the scarcity of water into a moment of
resourceful adaptation.

Finding Water in Urban Environments

Urban landscapes, despite their concrete facades, harbor
hidden water veins waiting to be tapped. This section
unfolds like a guide to urban exploration, revealing the
secret sources that lurk amidst the bustle of city life. From
storm drains to abandoned buildings, readers are guided
through the process of finding water in unexpected places.
It's not just about relying on the faucet; it's about cultivating
an urban survival mindset where every nook and cranny

becomes a potential oasis. In this section, water is not just a commodity; it's a treasure waiting to be discovered.

Extracting Water from Plants

Nature, in its silent generosity, provides a bounty of hydration in the very plants that surround us. This segment unfolds like a botanist's handbook, exploring the alchemical processes by which water can be extracted from plants. From succulent leaves to the moisture-laden roots, readers are guided through the subtle art of extracting water from the green tapestry of their surroundings. It transforms the act of admiring foliage into an active engagement with nature, where each plant becomes a potential wellspring of life-saving hydration.

Utilizing Emergency Water Packs

In the compact realm of emergency preparedness, water packs stand as modern-day ambrosia – a condensed source of hydration in times of need. This section is a guide to navigating the world of emergency water packs, exploring their diversity, shelf life, and strategic deployment. It's not just about stocking these packs; it's about understanding when and how to use them effectively. From the nuances of storage to the practicalities of carrying them on the go, readers are immersed in the art of integrating these compact reservoirs into their emergency water strategy.

Chapter 6

Community Water Preparedness

Amidst the challenges of water scarcity, Chapter 6 unfolds as a communal symphony, resonating with the collective heartbeat of shared preparedness. This chapter is not just about individual resilience; it's a call to cultivate a communal tapestry of water preparedness that interweaves the destinies of neighbors and friends. From the strategic blueprints of community water plans to the art of building a neighborhood water network, this chapter propels individuals beyond self-sufficiency into the realm of collaborative survival.

Establishing Community Water Plans

The foundation of community water preparedness lies in meticulous planning. This section is a guide to the orchestration of collective efforts, delving into the nuances of establishing community water plans. It's not just about individual households storing water; it's about creating a synchronized strategy that considers the vulnerabilities and strengths of the entire community. Readers are immersed in the intricacies of planning communal water sources, distribution points, and emergency response mechanisms. In this chapter, water preparedness becomes a shared endeavor where the strength of the community lies not just in the sum of its parts but in the cohesion of a well-thought-out plan.

Sharing Resources and Collaborative Solutions

A communal approach to water preparedness transcends mere planning; it delves into the ethos of sharing resources and collaborative problem-solving. This segment unfolds like a guide to fostering a sense of community resilience, where the surplus water in one household becomes a shared asset for all. From the logistics of communal storage to the establishment of mutual aid agreements, readers are guided through the art of building a network where water is not a private possession but a collective responsibility. It transforms water preparedness from an isolated endeavor into a communal ethic of sharing and support.

Building a Neighborhood Water Network

A neighborhood water network stands as the pinnacle of communal resilience – a dynamic tapestry that interconnects households, institutions, and emergency response mechanisms. This section is a journey into the practicalities of building such a network, exploring the technologies, communication strategies, and shared responsibilities that underpin its strength. It's not just about having neighbors; it's about forging connections that endure through the challenges of water scarcity. From the establishment of community water committees to the utilization of modern communication tools, readers are immersed in the transformative power of a neighborhood water network.

Chapter 7

Advanced Water Filtration Technologies

As the journey through water survival ventures into the realm of sophistication, Chapter 7 unfolds as an exploration of advanced water filtration technologies. It's a chapter that unveils the cutting-edge tools in the arsenal of water purification, where the art of filtration transcends basic methods. From the intricate workings of reverse osmosis systems to the transformative power of UV water purification and the innovation embedded in high-tech filtration devices, this chapter propels individuals into a realm where purity is not just a goal but a technological achievement.

Exploring Reverse Osmosis Systems

Reverse osmosis, a technological marvel in water purification, stands as the first pillar in this advanced exploration. This section unfolds like an engineer's manual, dissecting the mechanics of reverse osmosis systems. It's not just about forcing water through a membrane; it's about understanding the precision required to filter out contaminants at a microscopic level. Readers are immersed in the intricacies of membrane technology, pressure differentials, and the osmotic journey that water undertakes to emerge as a pristine elixir. In this chapter, reverse osmosis is not just a process; it's a technological ballet where water purity is achieved through a symphony of engineered precision.

UV Water Purification Methods

The power of the sun, harnessed through UV water purification methods, takes center stage in the quest for pure water. This segment is a journey into the transformative effects of ultraviolet light on waterborne pathogens. It's not just about exposing water to UV rays; it's about understanding the intricate dance where UV light disrupts the DNA of microorganisms, rendering them incapable of causing harm. Readers are guided through the nuances of UV purification, exploring its applications in portable devices, water treatment plants, and emergency scenarios. In this chapter, UV purification is not just a technology; it's a radiant force that annihilates threats on a microscopic scale.

High-Tech Filtration Devices

The landscape of high-tech filtration devices expands the repertoire of water purification into a realm of innovation and sophistication. This section unfolds like a showcase of technological wonders, exploring devices that utilize advanced materials, smart technologies, and multifaceted filtration mechanisms. It's not just about filtering water; it's about embracing the marriage of science and engineering to create devices that outperform conventional methods. From smart water bottles with embedded filters to portable purifiers with intricate filtration cartridges, readers are guided through the dazzling array of high-tech filtration options available. In this chapter, high-tech filtration is not just a luxury; it's a manifestation of human ingenuity crafting devices that elevate water purification to an art form.

Chapter 8

Maintaining Water Quality

As the voyage through water survival reaches its zenith, Chapter 8 emerges as a custodian's guide to maintaining water quality. It's a chapter that pivots from the acquisition and purification of water to the ongoing responsibility of ensuring its purity and safety. From the meticulous process of regular testing and monitoring to the strategic shield against contaminants and the foresight required for long-term water safety, this chapter transforms individuals from passive consumers of water to vigilant stewards of their liquid lifeline.

Regular Testing and Monitoring

The foundation of maintaining water quality lies in the diligent practice of regular testing and monitoring. This section unfolds like a laboratory expedition, guiding individuals through the nuances of water quality testing. It's not just about dipping a test strip into a sample; it's about understanding the parameters, interpreting results, and discerning the subtle signals that may hint at impending issues. Readers are immersed in the intricacies of pH levels, microbial content, and chemical composition, equipping them with the knowledge to transform water testing from a sporadic chore into a routine safeguarding practice.

Safeguarding Against Contaminants

The world around us is a dynamic tapestry of potential contaminants, and safeguarding against their intrusion becomes an art form in water maintenance. This segment is a journey into the strategic fortification of water sources against contaminants, exploring not just the removal of impurities but the prevention of their entry. It's not just about reacting to pollution events; it's about establishing barriers that shield water from potential threats. Readers are guided through the deployment of physical barriers, the implementation of best practices in waste disposal, and the cultivation of a mindset that places prevention at the forefront of water safety.

Addressing Long-Term Water Safety

The concept of maintaining water quality transcends the immediate; it extends into the realm of long-term water safety. This section unfolds like a planner's manual, guiding individuals through the foresight required to ensure a sustained supply of safe water. It's not just about addressing immediate concerns; it's about considering the potential evolution of risks over time. From the implications of climate change to the long-term effects of industrial activities, readers are immersed in the complexities of future-proofing their water sources. In this chapter, water safety is not just a reactive measure; it's a proactive strategy that anticipates and addresses potential challenges before they manifest.

CONCLUSION

Recap of Key Points

As we draw the curtains on this comprehensive exploration of water survival strategies, it's paramount to revisit the key points that weave through the tapestry of the preceding chapters. From the fundamental understanding of personal water needs to the exploration of diverse water sources and the mastery of purification techniques, each chapter has unfurled a layer in the intricate web of preparedness. The chapters guided readers through the art of DIY water collection, navigated the nuances of emergency water strategies, and ventured into the realms of advanced filtration technologies. From the individual to the communal, from the basic to the sophisticated, the key takeaway is the multifaceted nature of water preparedness.

Encouragement for Ongoing Preparedness

The journey through the pages of this survival compendium is not merely a one-time endeavor; it is a call to embrace ongoing preparedness as a way of life. Water, the essence of survival, demands perpetual vigilance and adaptability. The encouragement resonating from these pages is not a static cheer but a dynamic call to incorporate the insights gained into daily life. It's about turning the knowledge of water needs into a routine calculation, transforming the act of sourcing water into a deliberate exploration, and imbuing the mastery of purification methods into a practical skill. This chapter is not just a conclusion; it's an initiation into a mindset of perpetual readiness, where the lessons learned become an inherent part of the survival repertoire.

Additional Resources and References

In the quest for preparedness, knowledge is the compass that guides the way. As a parting gift, this section serves as a treasure trove of additional resources and references. It's not just about the culmination of knowledge within these pages; it's about extending the journey into deeper exploration. Readers are encouraged to delve into additional literature, online resources, and community forums that specialize in emergency preparedness and water survival. From books that explore the intricacies of water chemistry to forums where experiences are shared, the array of resources is diverse. This chapter unfolds like a map, pointing readers towards a continued journey of learning and discovery.

In conclusion, this survival guide is not just a manual; it's an empowering companion on the journey towards water resilience. From understanding the intricacies of personal water needs to fostering communal preparedness, from mastering purification techniques to embracing advanced filtration technologies, each chapter is a stepping stone in the quest for survival proficiency. As the reader closes this chapter, the journey doesn't end; it transforms into a mindset of perpetual readiness, where the wisdom gained becomes a beacon in navigating the unpredictable seas of crisis. The final encouragement is not just to absorb the knowledge but to live it – to turn preparedness into a way of life and water resilience into an enduring skill.